Address
4533 19th Ave NE
Seattle
WA 98105.
USA.

Mandy Whelan
July, 1988

WASHINGTON

WASHINGTON

✣

TOM & PAT LEESON

PAT O'HARA

✣

BOLTON

Designed by Fortunato Aglialoro
ISBN 0-920831-02-8
1 2 3 4 9 8 7 6
Printed in Hong Kong by Scanner Art Services, Inc., Toronto

STATE OF WASHINGTON
OFFICE OF THE GOVERNOR

OLYMPIA
98504-0413

BOOTH GARDNER
GOVERNOR

Dear Reader:

The people of Washington are very proud of their state and its beauty. Whether you visit the rolling hills of Eastern Washington or the immense peaks of the Cascade mountain range or take time to cruise the bays of Puget Sound, Washington's natural beauty will arouse your senses.

In the following pages you will witness firsthand Washington's beauty, dramatically captured by Tom & Pat Leeson and by Pat O'Hara in their photographic portrayal of our state.

I hope this book will enhance your knowledge of the splendor and beauty of Washington. And perhaps, you will visit Washington on your next vacation. We hope so!

Sincerely,

Booth Gardner
Governor

1 *(right)* Moon over Liberty Bell Mountain, early spring; near Washington Pass, on the scenic North Cascade Highway.

2 *(left)* Mount Rainier seen from Indian Henry's Hunting Ground; at 14,410 feet this is the highest peak in Washington.

3 Shilshole Bay, Seattle; some of the vessels that make Seattle the boating capital of the Northwest.

4 Wild rhododendron (*Rhododendron macrophyllum*), the state flower of Washington; prolific on the east side of the Olympic Peninsula and on the Hood Canal.

5 *(right)* Black-tail doe with twin fawns; the black-tail deer (*Odocoileus hemionus*) is the most common deer of Western Washington. Both mule deer and white-tail deer are also to be found within the borders of the state.

6 *(left)* Sunset on a windy evening, near Point of Arches on the North Washington coast. This is one of the newest additions to Olympic National Park, and its most northerly stretch of coastline.

7 Tum Tum Mountain at sunset, seen from the Mount Rainier Highway just a few miles southwest of the Paradise Visitor Center.

8 *(left)* Tulip fields near Mount Vernon and Anacortes; blooming time comes in April and May. Nearby fields host displays of daffodils and iris. Washington is one of the leading growers of all these bulb-flowers.

9 Built in 1893, the State Capitol at Olympia is famous for its classical architecture and the immaculate beauty of its gardens.

10 *(left)* Wheatfield near Dixie in Walla Walla County. Wheat is Washington's largest crop.

11 Yakima River between Cle Elum and Ellensburg. Old Highway 10 follows the Yakima for many miles and is designated as a scenic highway further east in the Yakima Canyon.

12 The Pacific Rhododendron (*Rhododendron macrophyllum*), the state flower, livens the forest of Western Washington with cascades of blossom in May and June.

13 *(right)* Maryhill Castle, overlooking the east end of the scenic Columbia River Gorge, is a baronial mansion built by the railroad magnate Samuel Hill in 1926 and now in use as an art museum.

14 A large, six-point bull elk. Found throughout the timbered regions of the state, the elk (*Cervus elaphus*) is one of Washington's principal big game species. Those found along the east slope of the Cascades and those in the Blue Mountains are descended from elk trapped in Yellowstone in the early 1900s.

15 *(right)* Black bear (*Ursus americanus*) with freshly caught salmon. Black bear are common throughout the forested parts of the state, where they live on a variety of plants and animals. Salmon, and steelhead trout, are favored not only by bears—they form the basis of one of Washington's major industries.

16 Rafting on the Wenatchee River.

17 *(right)* Early winter snows cover Hurricane Ridge and the higher peaks inside Olympic National Park.

18 *(left)* Glacier Peak, seen across Image Lake, in the Glacier Peak Wilderness.

19 One of a chain of volcanic peaks known as the Cascade Mountains, Mount St Helens lay dormant from 1897 until 1980, and the view of Mount St Helens across Spirit Lake was of a lovely symmetrical cone, very much like the view of Glacier Peak across Image Lake, seen opposite. Then Mount St Helens exploded. Here is the scene at sunset on 18 May 1980. Mount St Helens lost 1300 feet from its summit and devastated more than 235 square miles of forests and lakes (*see next page*).

20 *(left)* Mount St Helens National Volcanic Monument seen from the air. The destruction of the summit and the devastation around Spirit Lake are plain to see, even under the merciful shroud of snow, but already nature is on the mend again, with a vigorous renewal of plantlife.

21 Visible from many points in both Western and Eastern Washington, Mount Rainier, here seen from Pacific Crest Trail, has become a traditional symbol of the state.

22 *(left)* A Washington State Ferry prepares to dock in Bremerton as the sun goes down behind the Olympic Mountains.

23 Vine maple (*Acer circinatum*) brings fall color to the Douglas firs of the Cascades.

24 Fort Vancouver; blacksmith at work.

25 (*right*) Fort Vancouver, founded in 1824, was the western headquarters of the Hudson's Bay Company from 1825 until 1849. A hub of politics and fur-trading, the fort enclosed 22 buildings within its stockade. Today the stockade and several prominent buildings have been reconstructed on site and are open to the public.

26 (*left*) Washington produces one third of all the nation's apple crop. Late in September the eastern orchards are busy with the harvest. Red Delicious, like these in the Yakima Valley, have become synonymous with Washington State.

27 Farm near Sedro Woolley. The lowlands of Western Washington, especially the river valleys, have long been cultivated as farmlands.

28 North Head Lighthouse, one of two that guard the mouth of the Columbia River, began operating in 1898.

29 *(right)* The Tacoma Narrows Bridge with Mount Rainier in the background. Tacoma is one of the largest cities in Washington, a thriving port community where commerce is graced by glorious wilderness backdrops inland and on the Peninsula.

30 (*left*) With approximately 12 *feet* of precipitation a year, the Olympic Peninsula rain forest is the wettest spot in the 48 states. Green is the theme, since about ninety species of epiphytes (mosses, lichens and ferns) grow on the trees of the rain forest. It is also true that more specimens of gigantic trees of a wider variety of species grow here than grow in any other place in the world.

31 Farm near Goldendale. Along Highway 97 are many fine views of Mount Adams to the west. The nearby drive along the Klickitat River offers yet other beautiful views of the countryside.

S.F.&N.RY.
S.F.&N.RY.
S.F.&N.RY.
S.F.&N.RY.
S.F.&N.RY.
S.F.&N.RY.
Dr Pepper
7Up

32 (*left*) Site of the 1974 World's Fair, Riverfront Park in downtown Spokane still draws crowds on a summer weekend.

33 Hub of the 'Inland Empire' Spokane, on the eastern border of Washington, is one of the largest cities in the entire American Northwest and the cultural capital of its region, rich in museums and galleries and with scores of parks and gardens, among them the 100-acre site at the center of the city where the 1974 World's Fair was held.

34 (*left*) Prusik Peak with Gnome Tarn, in the Alpine Lakes Wilderness, Washington is fortunate in having several million acres of federal lands included within the National Wilderness Preservation System. Just east of Seattle, along the crest of the Cascades, lies the beautiful Alpine Lakes Wilderness, between the Stevens and Snoqualmie Passes.

35 Built in 1921 to commemorate more than a hundred years of peace between the United States and Canada, the Peace Arch stands 67 feet high at the border near Blaine. Much of the cost of construction was met by donations from schoolchildren in British Columbia and Washington.

36 *(left)* Palouse Falls State Park, Whitman County. The Palouse tumbles over a basalt headwall on its way to join the Snake River in Eastern Washington. In spring such wildflowers as balsamroots line the upper levels of the cliffs.

37 Part of the Wild and Scenic River System, the upper Skagit is a good location for bald eagles from November through March, the birds being attracted to the river by its large runs of anadromous fish—which also draw the human fishermen as well.

38 Matia Island, San Juan Archipelago, just north of Orcas Island. Matia is managed as two parts, one a State Park and the other a National Wildlife Refuge.

39 (*right*) Manastash Creek, on the eastern slope of the Cascades. Cottonwood trees and red-osier dogwood line the rivers and streams of Eastern Washington. In fall the watercourses are decorated by brilliant displays of red and yellow leaves.

40 Farmers haying near Sequim. An old pastoral farming community, the Sequim area has in recent years become a retirement haven as more and more people come to enjoy its beauty, its mild climate and its only moderate rainfall. Sequim lies in the rain-shadow of the Olympic Mountains.

41 (*right*) A rainbow heralds the end of a brief summer shower over a hayfield near Nespelem.

42 Irrigation circles north of the Columbia River near Paterson dominate land that once was covered with sage and prickly pear. Given only twelve inches or less of rain in a year, irrigation is needed for there to be corn, potatoes and alfalfa in this sundrenched environment.

43 *(right)* The 'Blue Bridge' across the lower Columbia River joins Kennewick and Pasco. These two cities, together with Richland the largest of three, make up the community generally known as the 'Tri-Cities'.

Ski Tavern
Der Sportsmann

44 (*left*) Designed in Bavarian style, Leavenworth on US Highway 2 is known for its autumn leaf festival in October. Nearby Icicle Creek converges with the Wenatchee River which in turn feeds the mighty Columbia downstream.

45 A clown at the Ellensburg Rodeo distracts a bucking bull and so gives the bull's fallen rider a chance to get safely out of the arena. Rodeos are popular weekend entertainment in Eastern Washington from early May through Labor Day weekend.

46 Daniel Stuart, a 4-H young man, gives a little last-minute grooming to his Suffolk Cross ewe, prior to the judging at the Clark County Fair.

47 *(right)* One of the many excellent produce shops that enliven the Pike Street Market area of downtown Seattle.

PRODUCE
Made to Order
BROCCOLI
59
SWEET CARROTS
VINE RIPE TOMATOES
69
GREEN LEAF
59
RED LEAF
LETTUCE
59
LEEKS
125
40
BABY DILL
FRESH BASIL
195
BABY ARTICHOKES
99¢
POUND
CHERRY TOMATOES
75¢
BASK.
FRESH MINT
SWEET ANISE
$1.00
EACH
FRESH CILANTRO
CALBERI
CALIFORNIA STRAWBERRIES

48 Log trucks wait in Grays Harbor, Port of Hoquiam, to load their cargo onto a log ship bound for the Far East. Washington exports around two billion board-feet of timber every year, with most of it going to Japan but an increasing amount to China.

49 (*right*) Log trucks waiting at a log dump near Hoquiam. The oldest industry in Washington, logging was a major industry in the western region even before the state was formed. The first sawmill began operating as early as 1825. Washington accounts for about one-fifth of all the Douglas fir harvest in the nation.

50 Mount Shuksan, seen from Highwood Lake, North Cascades National Park. October is especially beautiful here, when the huckleberry and the mountain ash are in their fall colors.

51 (*right*) Dairy farm, Cle Elum. Almost half the land in Washington is farmland and of that one half is under crops and the other half given to pasture and small woodlots.

52 Hastings House, Port Townsend. Many Victorian homes have been restored in Port Townsend and a number are open for scheduled public tours. This architecture recalls the late 1870s when Port Townsend was one of the most elegant and busiest port cities of the whole West Coast.

53 (*right*) Washington State University, Pullman. Especially regarded for the importance of its agricultural faculties, this large university is set in the picturesque and fertile Palouse Hills in Eastern Washington.

54 (*left*) Rainbow over sagebrush-covered hills in Southeastern Washington. Much of the eastern region looked like this before farming became a way of life.

55 Hunting, fishing and camping are major recreational activities. Here an outfitter leads a packhorse through the aspens on a fall ride in Northeastern Washington.

56 Ranch in Kittitas Valley, Eastern Washington. About 50 miles east of Snoqualmie Pass along the Interstate 90 corridor lies picturesque Kittitas Valley. Lying in the rain shadow of the Cascades, this is an area of hot summers and cold winters. Farming and ranching form the economic base of Kittitas Valley.

57 (*right*) Kettle Mountains, North Central Washington. A proposed 'wilderness area', this is an old mountain range of rolling ridges. Over the centuries fire has opened a beautiful series of meadows along the Kettle Divide and it is home for many mule deer in the summer months.

58 & 59 (*right*) Agriculture is the number one industry of Washington and wheat is the number one crop. The sale of Washington wheat brings in over 650 million dollars a year. Here we see a wheat-pile outside a grain elevator in Eastern Washington and (*right*) harvest time in Palouse country.

60 (*left*) Cape Disappointment Lighthouse at the mouth of the Columbia River. The Cape gets its name from an English sea-captain, John Meares, who failed to locate the Columbia in 1788. In November 1805 this is where Lewis and Clark first set eyes on the Pacific. The light came into operation on 15 October 1856 and since then has guided ships through one of the most treacherous passages on the entire West Coast.

61 A spring shower passes briefly over the Yakima River Valley near Prosser. Irrigation, not rainfall, makes this valley one of the most productive fruit and vegetable areas of Washington.

62 One of the longest natural landspits in the world, Dungeness Spit stretches five and one-half miles into the Strait of Juan de Fuca near Sequim. Much of the Spit was designated as a National Wildlife Refuge in 1915, and it is a good place to see loons, grebes, sea-ducks and shorebirds.

63 (*right*) Several hundred thousand shorebirds pause each spring on the rich, tidal mud-flats of Grays Harbor, to gorge themselves on a variety of saltwater organisms, before migrating onward to start another nesting season in the Arctic.

64 (*left*) Dungeness Flat and the Olympic Mountains. With a mountainous backdrop to the south and the Strait of Juan de Fuca to the north, Dungeness Flat is one of the most scenic agricultural settings on the Olympic Peninsula.

65 Big Beaver Valley, Ross Lake National Recreation Area. Adjacent to the North Cascades National Park, Big Beaver Valley hosts one of the most impressive stands of Western Red Cedar left in Washington State. A popular backpacking loop starts at Ross Lake, goes up Little Beaver Valley, over Beaver Pass and down Big Beaver Valley, returning to Ross Lake after a 27-mile trip.

66 (*left*) Cathlamet is a small harbor on the lower Columbia. Here the Hume brothers opened the first salmon cannery in Washington.

67 Fishing boats moored on a small slough on Puget Island in the lower Columbia River. Fishing used to be a way of life along the lower river but declining salmon runs have forced most fishermen into other occupations.

68 Dancing at the Indian encampment, Ellensburg Rodeo.

69 (*right*) Salmon bake during the annual Makah Days celebration at Neah Bay in August.

70 Canoeing on the calm waters of Fish Lake at the border of the magnificent Alpine Lakes Wilderness.

71 (*right*) Whitman Mission National Historic Site, where Dr. Marcus Whitman established his mission 1836-47, near Walla Walla, on the Old Oregon Trail. Whitman College was founded in 1859 and named in his memory. Walla Walla is now the largest city in Southeastern Washington.

72 (*left*) The green of early spring tinges the rolling grain fields of the Palouse country. In some places the dark, rich topsoil is as much as 200 feet deep. The name Palouse derives from the native people who originally inhabited the area.

73 Columbia River Gorge near Lyle. The Gorge stretches eastward from near the town of Washougal to the area of the John Day Dam, through one of the most scenic parts of Washington. One third of all the hydroelectric power in the 48 states is to be drawn from the Columbia River system.

74 (*left*) Boeing 767s in the final stages of assembly at the Boeing plant near Everett. The Boeing Company, founded in 1916, has manufactured more commercial jetliners than any other company in the world. Boeing is the largest employer in the State and the Everett plant also has the distinction of being the largest building in the world, nearly half a mile in length.

75 Grand Coulee Dam is the largest concrete structure and, with the recent addition of a third powerhouse, the largest hydroelectric installation in the world. Built between 1933 and 1942 the dam is over 4000 feet across and 550 feet high. Generating capacity—over nine million kilowatts.

76 Male American Goldfinch (*Carduelis tristis*) with a favorite delicacy—wild thistle seeds. This is the state bird of Washington.

77 (*right*) A grove of Western Red-Cedar, Ross Lake National Recreation Area. Mature stands of these noble trees were common in the early days of Washington State, but through persistent cutting their numbers have steadily diminished over the past century. A few virgin stands are preserved, as a window into the past of these once extensive forests.

78 (*left*) A wheat farm at the foot of the Blue Mountains near Walla Walla. This region is famous not only for its yield of grain but also for Walla Walla sweet onions.

79 Looking west from Grand Peak across the mountain ridges of Olympic National Park. This wilderness of the interior is reached by a network of nearly 600 miles of trails for horse and foot.

80 Bald eagle (*Haliacetus levcocephalus*) landing in a treetop. Bald eagles have been on the increase in Washington lately, thanks to better protection and the banning of DDT. Winter is the best time to see these birds as many migrate south from Alaska and Canada to winter along the coast and to feed on spawned-out salmon along the rivers.

81 (*right*) Mountain lion (*Felis concolor*) crossing an alpine meadow in the Cascade Mountains. The mountain lion is America's largest cat, measuring as much as eight feet in length and 250 pounds in weight. Yet it is extremely shy and rarely seen.

82 *(left)* At 7,965 feet the tallest peak on the Olympic Peninsula, Mount Olympus looks down on a subalpine meadow covered with avalanche lilies. Olympic National Park is roughly 900,000 acres in extent. It was established in 1938 and remains mostly unchanged in its natural state.

83 Avalanche lilies (*Erythronium montanum*) or fawn lilies make one of the showiest alpine floral displays. They usually bloom soon after the snow melts, in June and July: two of the best places to see them in profusion are Mount Rainier National Park and Olympic National Park.

84 (*left*) Sunset along the Strait of Juan de Fuca. 'The Strait' as it is known locally, is a major shipping route to the harbors of Puget Sound.

85 Pleasure-boats on Lake Union, skyscraper profile of downtown Seattle shines in the sunset and the 76-storey Columbia Center towers beyond.

86 Aerial view of Olympic National Park, with Mount Olympus in early morning light.

87 (*right*) Puffin Island, San Juan Archipelago, with Mount Baker in the distance. Many of the San Juan Islands are small outcrops in the Strait of Juan de Fuca. Many are protected by the National Wildlife Refuge System. Marine mammals, birds and tidepool life depend upon these islands for their existence.

88 (*left*) Sailboats are a favorite pastime of Washingtonians who live in the Puget Sound area.

89 One of the most spectacular views of the San Juan Islands is to be seen at Sunset from Chuckanut Mountain south of Bellingham. Just follow the good gravel road. . . .

90 Sunset near Ruby Beach, Olympic National Park, on Washington's northern coastline.